RIEP CHART
PRESENTS:

FUNDAMENTALS OF ELECTRICITY

#	R	I	E	P
R_1	1	2	2	4
R_2	2	2	4	8
R_3	3	2	6	12
R_T	6	2	12	24

AKA

HAVING FUN WITH CIRCUIT ANALYSIS

<u>*NO RETURNS / NO REFUNDS ON THIS WORKBOOK*</u>

Contents

Author's Note

This workbook was designed to build upon one idea at a time to maximize student learning, with the aid of your instructor. It is not a regular basic electricity book. There are enough electricity books out there, that we certainly do not need another one. Although this material is best suited for the classroom work during lab, it can be used by any student wanting to learn. What makes this material so difficult to understand is the fact that one learning style will not work effectively the whole way through this process, unlike most other subjects. The different concepts in electricity are best developed by implementing a couple of different learning styles.

Please Note; You do **NOT** want to learn the practical (labs, meters, live circuits, etc.) electricity on your own. It really can kill, or injure you if you do not know and apply all the safety rules and practices. And no, the meters and lab equipment referenced in this workbook are not included.

It should also be noted that the solutions to the different problems are **NOT** inside this workbook. That is by design so the instructor can control the formal process. The answers along with the recommended teaching points with the best practices for circuit analysis can be found inside the instructor workbook, "**Secrets of the RIEP Chart**". One last thing, the lab equipment and materials mentioned and needed for the completion of the worksheets are not included with this workbook.

INTRODUCTION

Hello,

Welcome to the, "**RIEP Chart**" circuit analysis made easy. This being college you should have already read your required textbooks, twice, before the first day of class. I should add that not everything you read will make sense to you, but you will conceptually understand where you will be at the end of the semester. The instructor's job is to make electricity come alive (pun intended) and to answer your questions. Let there be no doubt that electricity is as difficult to learn as a second language. However, if you come to class each and every meeting ready to learn, you and your instructor will together raise your understanding of electricity.

No matter what your background in electricity is, you will enjoy this class. So let's get started. Before you can build your first circuit, you must learn some new terms and how to operate your meter. The 100 words listed under the words to define will need to be fully understood before you can really understand basic electricity.

Every instructor teaches their class utilizing slightly different processes. Some instructors will use more worksheets than others, so do not think you will do all the worksheets in this workbook, which is up to the instructor's discretion. Therefore, it is highly recommended that each student pay close attention to the directions and the additional worksheets identified by the instructor. With that said however, please do not try to get ahead of the class by finishing this workbook before the instructor can cover how they want you to fill it in.

This workbook that you now have might seem overwhelming to you if you have never had an electricity class before. The instructor will be using this during class time, before and during labs, so do not worry that you might not understand it all. If you show up at all the class meetings and want to learn, you will.

Every one of these pages will introduce a new subject or another way of looking at a circuit. Every book you read about electricity will try to make you understand their way of labeling things. Truth be told sometimes we have the ability to use six different labels for the same thing. That in itself can be quite confusing. We will endeavor to transcend the math to allow you to fully understand the introduction to, and the general concepts of this thing called, **"BASIC CIRCUIT ANALYSIS."**

WORDS TO DEFINE

1. Absolute Zero.
2. Alternating Current.
3. Alternation
4. Alternator.
5. Ammeter.
6. Ampere.
7. Ampere turn.
8. Amp-Hour.
9. Anode.
10. Apparent Power.
11. Armature.
12. Armature reaction.
13. Atom.
14. Attenuation.
15. Avalanche Voltage.
16. Average Value.
17. Base.
18. Battery.
19. Bias.
20. Bimetallic Element.
21. Bonding.
22. Breakdown Voltage.
23. Capacitance.
24. Capacitive Reactance.
25. Capacitor.
26. Cathode.
27. Circuit.
28. Circular Mil.
29. Closed Circuit Voltage.
30. Coil.
31. Commutator.
32. Compound Winding.
33. Conductor.
34. Continuity.
35. Copper Loss.
36. Coulomb.
37. Counter EMF.
38. Current.
39. Delta Connection.
40. Dielectric.
41. Direct Current.
42. Domain.
43. Eddy Currents.
44. Effective Value.
45. Electricity.
46. Electrolyte.
47. Electromagnetic Induction.
48. Electromotive Force.
49. Electron.
50. Farad.
51. Frequency.
52. Generator.
53. Growler.
54. Henry.
55. Hertz.
56. Horse Power.
57. Hysteresis Losses.
58. Impedance.
59. Inductance.
60. Inductive Reactance.
61. Inductor.
62. Insulator.
63. Interpoles.
64. Inverter.
65. Ion.
66. Magnet.
67. Magneto Motive Force.
68. Megger.
69. Mho.
70. Neutron.
71. Ohm.
72. Ohm's Law.
73. Open Circuit Voltage.
74. Peak Inverse Voltage.
75. Peak Value.
76. Piezoelectric Effect.
77. Polarity.
78. Potential.
79. Power.
80. Power Factor.
81. Primary Cell.
82. Proton.
83. Reactive Power.
84. Real Power.
85. Reluctance.
86. Resonance.
87. Root Mean Square.
88. Secondary Cell.
89. Shunt.
90. Sine Wave.
91. Time Constant.
92. Transformer.
93. True Power.
94. Valence Electron.
95. Volt.
96. Volt Amp Reactive.
97. Voltage Divider.
98. Voltmeter.
99. Watt.
100. Watt Hour.

RESISTOR COLOR CODE # 1

RESISTOR NUMBER	FIRST COLOR	SECOND COLOR	THIRD COLOR	FOURTH COLOR	CODED VALUE	TOLERANCE (%)	MEASURED VALUE	GOOD YES / NO
R_1					1700	20%	N/A	N/A
R_2					2300	10%	N/A	N/A
R_3					900	5%	N/A	N/A
R_4					57	5%	N/A	N/A
R_5					4.7	10%	N/A	N/A
R_6					0.23	20%	N/A	N/A
R_7					75	5%	N/A	N/A
R_8					56K	10%	N/A	N/A
R_9					37M	5%	N/A	N/A
R_{10}					97K	10%	N/A	N/A
R_{11}					25	20%	N/A	N/A
R_{12}					2.5K	20%	N/A	N/A
R_{13}					6K	10%	N/A	N/A
R_{14}					8.7M	5%	N/A	N/A
R_{15}					100K	20%	N/A	N/A
R_{16}					30	5%	N/A	N/A
R_{17}					50M	5%	N/A	N/A
R_{18}					.9K	20%	N/A	N/A
R_{19}					500	10%	N/A	N/A
R_{20}					1K	5%	N/A	N/A

RESISTOR COLOR CODE # 2

RESISTOR NUMBER	FIRST COLOR	SECOND COLOR	THIRD COLOR	FOURTH COLOR	CODED VALUE	TOLERANCE (%)	MEASURED VALUE	GOOD YES / NO
R_1								
R_2								
R_3								
R_4								
R_5								
R_6								
R_7								
R_8								
R_9								
R_{10}								
R_{11}								
R_{12}								
R_{13}								
R_{14}								
R_{15}								
R_{16}								
R_{17}								
R_{18}								
R_{19}								
R_{20}								

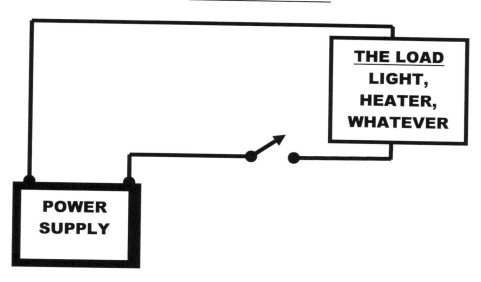

R#	R	I	E	P
R_1	1	5		
R_2	2	4		
R_3		7	14	
R_4			9	27
R_5		5	30	
R_6	6			96
R_7	6			150
R_8	7			63
R_9	14			56
R_{10}	5			180

RIEP CHART # 2

#	R	I	E	P
R_1	5	6		
R_2		2	200	
R_3		5	250	
R_4			60	240
R_5	10			1000
R_6	5			500
R_7	20	3		
R_8			100	400
R_9	17			68
R_{10}		6	66	
R_{11}			20	100
R_{12}		4	144	
R_{13}	9			576
R_{14}	1.5K	2ma		
R_{15}	2.2K		11	
R_{16}	4.7K			47
R_{17}	35K			87.5
R_{18}	3.3K	10ma		
R_{19}	192K			192mw
R_{20}	1K	20ma		
R_{21}	25		75	

DC SERIES - PROBLEM # 1

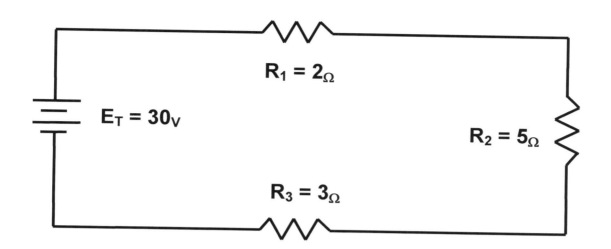

#	R	I	E	P
R_1	2_Ω			
R_2	5_Ω			
R_3	3_Ω			
R_T			30_V	

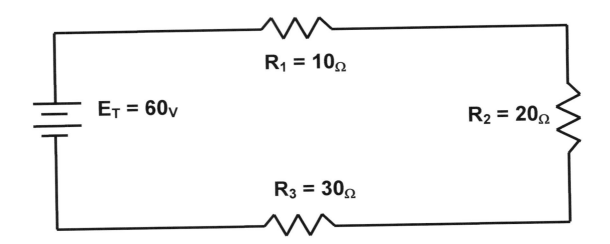

R₁ = 10Ω

E_T = 60_V

R₂ = 20Ω

R₃ = 30Ω

#	R	I	E	P
R₁	10Ω			
R₂	20Ω			
R₃	30Ω			
R_T			60_V	

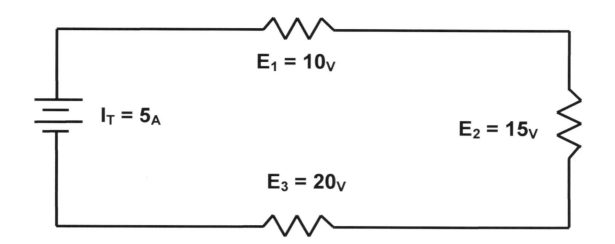

$E_1 = 10_V$

$I_T = 5_A$

$E_2 = 15_V$

$E_3 = 20_V$

#	R	I	E	P
R_1			10$_V$	
R_2			15$_V$	
R_3			20$_V$	
R_T		5$_A$		

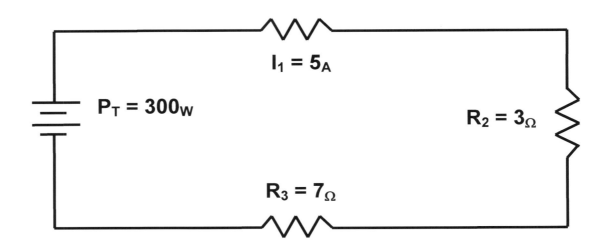

$I_1 = 5_A$

$P_T = 300_W$

$R_2 = 3_\Omega$

$R_3 = 7_\Omega$

#	R	I	E	P
R_1		5_A		
R_2	3_Ω			
R_3	7_Ω			
R_T				300_W

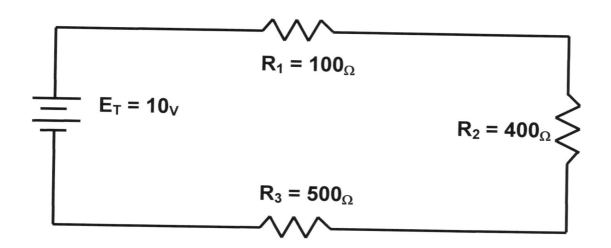

#	R	I	E	P
R_1	100$_\Omega$			
R_2	400$_\Omega$			
R_3	500$_\Omega$			
R_T			10$_V$	

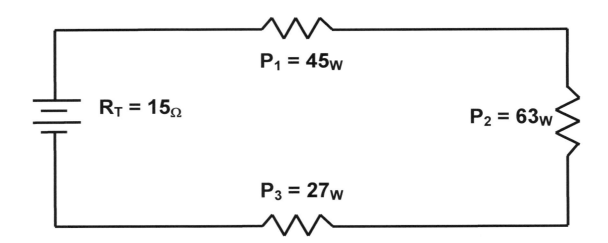

#	R	I	E	P
R_1				45_W
R_2				63_W
R_3				27_W
R_T	15_Ω			

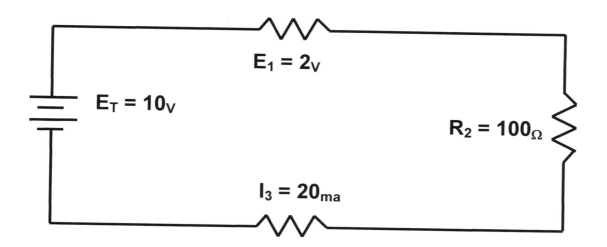

#	R	I	E	P
R_1			2_V	
R_2	100_Ω			
R_3		.02		
R_T			10_V	

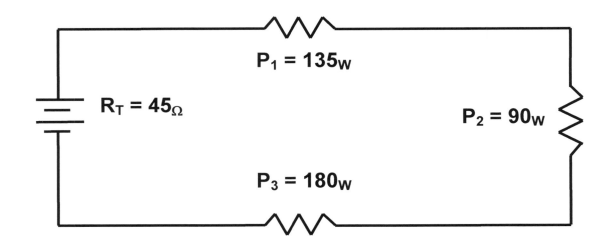

$P_1 = 135_W$

$R_T = 45_\Omega$

$P_2 = 90_W$

$P_3 = 180_W$

#	R	I	E	P
R_1				135$_W$
R_2				90$_W$
R_3				180$_W$
R_T	45$_\Omega$			

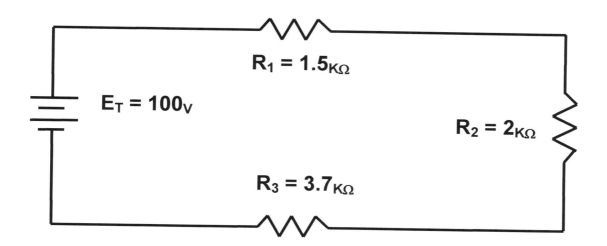

#	R	I	E	P
R_1	1500$_\Omega$			
R_2	2000$_\Omega$			
R_3	3700$_\Omega$			
R_T			100$_V$	

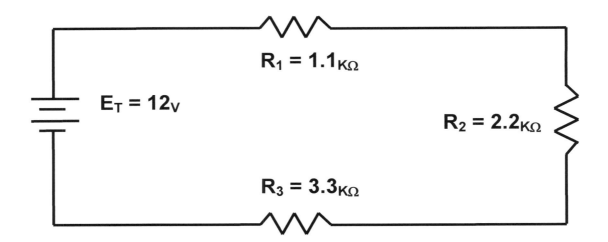

$E_T = 12_V$

$R_1 = 1.1_{K\Omega}$

$R_2 = 2.2_{K\Omega}$

$R_3 = 3.3_{K\Omega}$

#	R	I	E	P
R_1	$1.1_{K\Omega}$			
R_2	$2.2_{K\Omega}$			
R_3	$3.3_{K\Omega}$			
R_T			12_V	

DC SERIES CIRCUIT - LAB WORKSHEET

INSTRUCTIONS:

1. Refer to a standard color code chart and determine the value of each resistor. Record the value of each.

   R₁ = _____   **R₂** = _____   **R₃** = _____

2. Using the ohmmeter function of a multimeter, measure the resistance of each resistor in # 1.

   R₁ = _____   **R₂** = _____   **R₃** = _____

3. Are all resistors within their tolerances?

   R₁ = YES / NO   **R₂** = YES / NO   **R₃** = YES / NO

R₁ = _____ Ω

E_T = _____

R₂ = _____ Ω

R₃ = _____ Ω

FIGURE # 1
Note: <u>All resistance values **MUST** be measured values - from a meter.</u>

4. Compute (with a calculator, think ***RIEP chart***) the total resistance in the circuit of figure 1.

   Record = _____

5. Compute the total current flow in the circuit of figure 1.

   Record = _____

6. Compute the voltage drop across each of the resistors in figure 1.

   E₁ = _____   **E₂** = _____   **E₃** = _____

7. <u>CAUTION:</u> Make sure the training equipment is turned **OFF**. If the trainer is not connected to the 115 volts AC power source (standard outlet), call instructor!

8. Using the resistors and wire jumpers provided, connect the three (3) resistors as shown in figure 1.

9. Using the ohmmeter function of the multimeter, measure the total resistance of the circuit in figure 1.

Total Resistance = _____, is this value the same as the number in step #4? YES / NO

10. Does the total value obtained in step #2 if added together, equal the value obtained in step #9?

Explain: _____

_____

_____

11. Have **INSTRUCTOR** check the Multimeter Trainer Panel before you continue!!!

12. Place the Multimeter Trainer Panel power switch in the on position.
    NOTE; IF YOU SMELL SMOKE PLEASE LET THE INSTRUCTOR KNOW.

13. Using the voltmeter function of the multimeter, measure the voltage drop (IR drop) across each of the resistors in the circuit.

E_1 = _____ E_2 = _____ E_3 = _____

14. Using the voltmeter function of the multimeter, measure the voltage drop (IR drop) across the closed **DC** circuit voltage (full-wave Rectifier Bridge).
    Total Voltage = E_T = _____

15. Does the voltage computed in step #6 equal the voltage measured in step #13 when all three are added together? YES / NO. Explain:

_____

_____

_____

16. Check with **INSTRUCTOR** before proceeding.

17. Using the ammeter function of the multimeter, measure the current in the circuit at the following points:
    a. Between the positive power supply and R_1 = _____
    b. Between R_1 and R_2 = _____
    c. Between R_2 and R_3 = _____
    d. Between R_3 and the negative power supply = _____
    e. Is the current the same at each point? = YES / NO
    f. Explain: _____

_____

18. Place all equipment in the off position and put away all jumpers (good housekeeping).

19. What precaution should be taken when using an ohmmeter?

_____

_____

20. When using a meter to measure an unknown value, what range should be used first and why?

_____

_____

21. In what part of a meter range (analog) is the scale the most accurate? _____

22. How is a volt meter connected to a circuit? _____

DC LAMP - LAB WORKSHEET

INSTRUCTIONS:

1. Measure and record resistance of incandescent lamp = _____.

2. Using Ohm's Law, calculate the lamp current based on a voltage of _____v,

    Record, I_T = _____.

3. Connect the lamp to the same voltage in step # 2.

4. Connect the ammeter (in series) with the lamp, and measure, I_T = _____.

5. Using Ohm's Law, calculate the lamp resistance using the current flow in step #4.

    Record, R_T = _____

6. Is the current (I) you measured in step #4 the same as you calculated in step #2?

    STEP # 2, I_T = _____   YES / NO   STEP # 4, I_T = _____

    a) Explain the results (why) in both quantitative and qualitative terms.

    _____

    _____

    _____

    _____

    _____

    _____

7. Place all equipment in the off position.

8. How is an ammeter connected to a circuit? _____

9. How do you describe a Series circuit and what are the four rules?

    _____

    _____

    a) Rule #1. R_T = _____

    b) Rule #2. I_T = _____

    c) Rule #3. E_T = _____

    d) Rule #4. P_T = _____

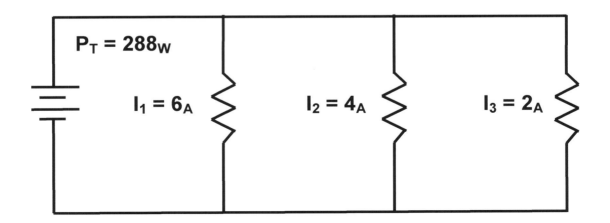

#	R	I	E	P
R_1		6ᴀ		
R_2		4ᴀ		
R_3		2ᴀ		
R_T				288ᴡ

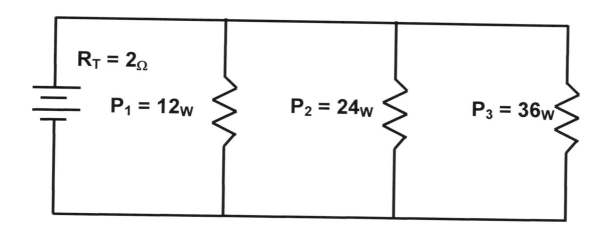

#	R	I	E	P
R_1				12_W
R_2				24_W
R_3				36_W
R_T	2_Ω			

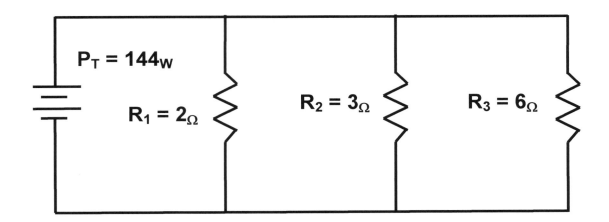

#	R	I	E	P
R_1	2_Ω			
R_2	3_Ω			
R_3	6_Ω			
R_T				144_W

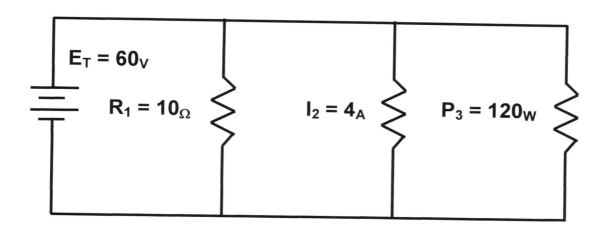

#	R	I	E	P
R₁	⓪10$_\Omega$			
R₂		④4$_A$		
R₃				⓪120$_W$
R$_T$			⓪60$_V$	

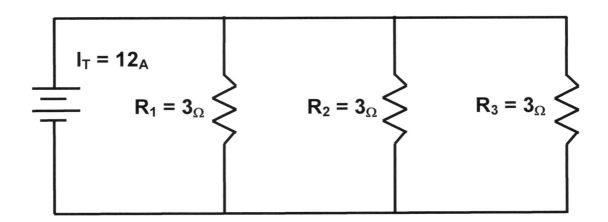

#	R	I	E	P
R₁	3Ω			
R₂	3Ω			
R₃	3Ω			
R_T		12_A		

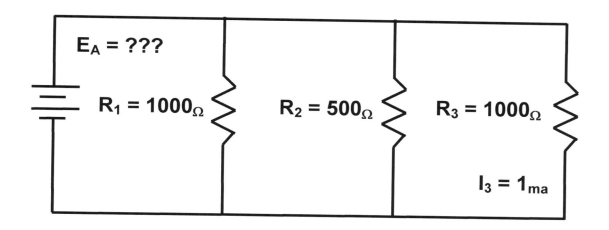

#	R	I	E	P
R_1	1000$_\Omega$			
R_2	500$_\Omega$			
R_3	1000$_\Omega$	.001$_A$		
R_T				

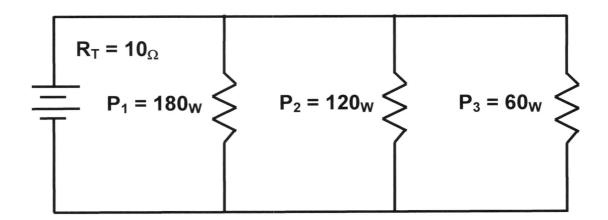

$R_T = 10_\Omega$

$P_1 = 180_W$ $P_2 = 120_W$ $P_3 = 60_W$

#	R	I	E	P
R_1				180$_W$
R_2				120$_W$
R_3				60$_W$
R_T	10$_\Omega$			

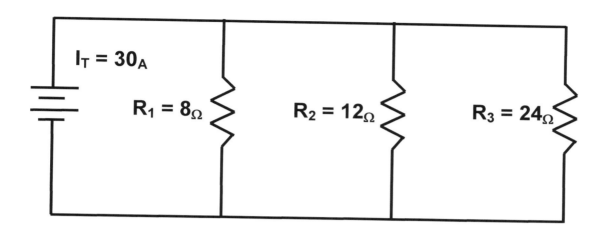

#	R	I	E	P
R_1	8_Ω			
R_2	12_Ω			
R_3	24_Ω			
R_T		30_A		

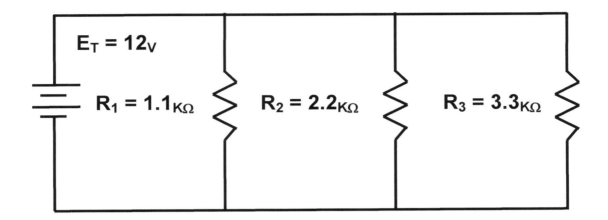

#	R	I	E	P
R_1	1.1$_{K\Omega}$			
R_2	2.2$_{K\Omega}$			
R_3	3.3$_{K\Omega}$			
R_T			12$_V$	

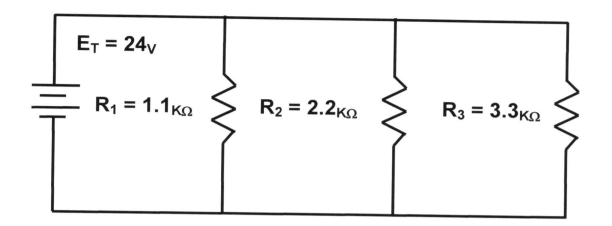

#	R	I	E	P
R₁	1.1 KΩ			
R₂	2.2 KΩ			
R₃	3.3 KΩ			
R_T			24 V	

DC PARALLEL CIRCUIT - LAB WORKSHEET

INSTRUCTIONS:

1. Refer to a standard color code chart and determine the value of each resistor. Record the values.

   R_1 = _____ R_2 = _____ R_3 = _____

2. Using the ohmmeter function of a multimeter, measure the resistance of each resistor in figure # 1.

   R_1 = _____ R_2 = _____ R_3 = _____

3. Are all resistors within their tolerances?

   R_1 = YES / NO R_2 = YES / NO R_3 = YES / NO

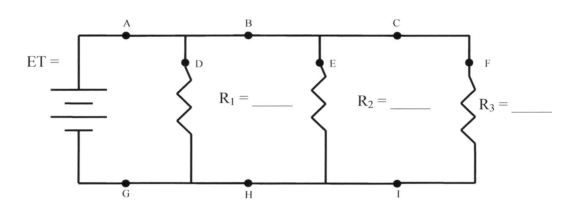

Figure # 1

4. Compute the R_T of the circuit in figure # 1. Record = _____

5. What should the **IR drop** be across the total resistance? _____

6. What should the **IR drop** be across each of the resistors?

   E_1 = _____ E_2 = _____ E_3 = _____

7. Compute the total current (I_T) in figure 1. Record = _____

8. Compute the current in each leg (branch) in the circuit.

   I_1 = _____ I_2 = _____ I_3 = _____

9. CAUTION: Make sure the training equipment is turned off. If the trainer is Not connected to the 115 volt AC power source, call **INSTRUCTOR**!

10. Using only the wire jumpers and the resistors provided, connect the three (3) resistors in parallel as shown in figure # 1.

11. Call **INSTRUCTOR** to look at trainer. **DO NOT TURN ON!**

12. Using the ohmmeter function measure the total resistance in the circuit of figure #1.

    R_T = _____

13. Is the value obtained in step #12 greater or less than the value obtained in step #4?

    Explain = _____

    _____

    _____

14. NOW place the Multimeter trainer Panel switch in the **ON** position.

15. Measure the voltage drop across the circuit. E_T = _____

16. Measure the voltage drop (**IR**) across each resistor.

    E_1 = _____ E_2 = _____ E_3 = _____

17. Is the value obtained in step #15, the same as each of the values obtained in step #16?

    Explain. _____

    _____

18. Using the ammeter function of the multimeter, measure the circuit at points A through I.

    A = _____   **B** = _____   **C** = _____   **D** = _____

    E = _____   **F** = _____   **G** = _____   **H** = _____

    I = _____

19. Place the trainer power switch in the off position, remove all jumpers.

20. How is an ammeter connected to a circuit? _____

21. Why does the R_T (resistance total) go down every time you add a resistor in a parallel circuit? Explain. _____

    _____

22. Why can there be a difference between the **OCV** (Open Circuit Voltage) and the **CCV** (Closed Circuit Voltage) for any given circuit (power supply)?

    _____

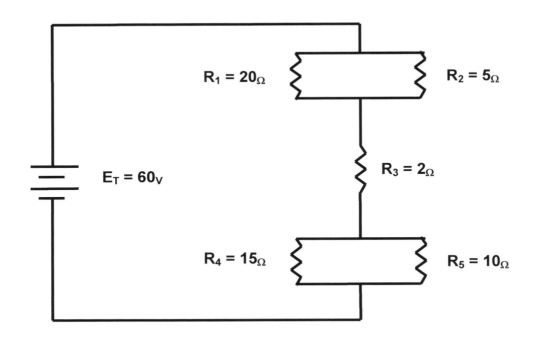

#	R	I	E	P
R_1	20$_\Omega$			
R_2	5$_\Omega$			
R_3	2$_\Omega$			
R_4	15$_\Omega$			
R_5	10$_\Omega$			
R_T			60$_V$	

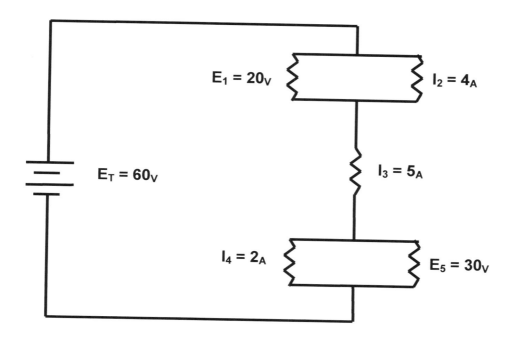

#	R	I	E	P
R₁			20ᵥ	
R₂		4ₐ		
R₃		5ₐ		
R₄		2ₐ		
R₅			30ᵥ	
Rₜ			60ᵥ	

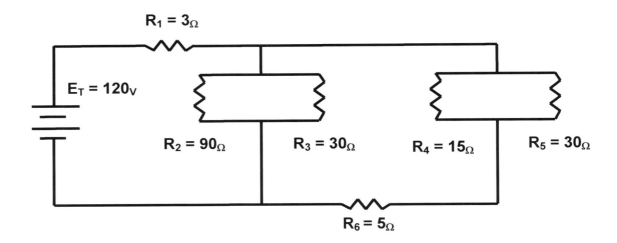

#	R	I	E	P
R_1	3_Ω			
R_2	90_Ω			
R_3	30_Ω			
R_4	15_Ω			
R_5	30_Ω			
R_6	5_Ω			
R_T			120_V	

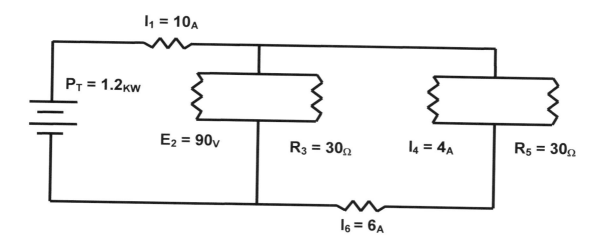

#	R	I	E	P
R_1		10₍A₎		
R_2			90₍V₎	
R_3	30₍Ω₎			
R_4		4₍A₎		
R_5	30₍Ω₎			
R_6		6₍A₎		
R_T				1200₍W₎

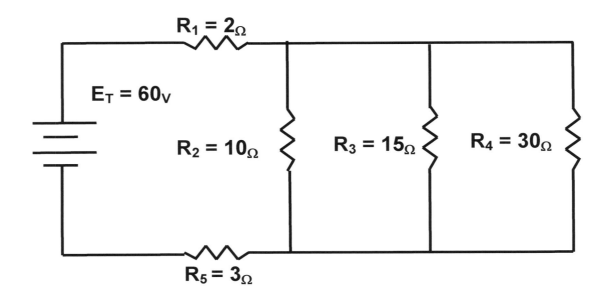

#	R	I	E	P
R_1	2_Ω			
R_2	10_Ω			
R_3	15_Ω			
R_4	30_Ω			
R_5	3_Ω			
R_T			60_V	

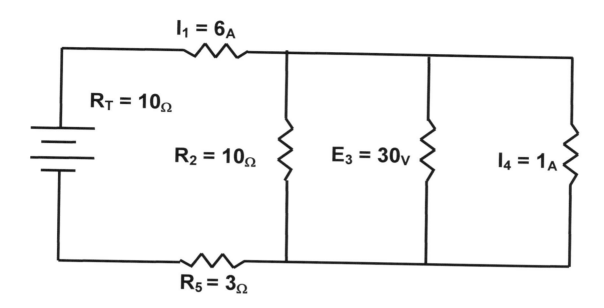

$I_1 = 6_A$

$R_T = 10_\Omega$

$R_2 = 10_\Omega$

$E_3 = 30_V$

$I_4 = 1_A$

$R_5 = 3_\Omega$

#	R	I	E	P
R_1		6_A		
R_2	10_Ω			
R_3			30_V	
R_4		1_A		
R_5	3_Ω			
R_T	10_Ω			

DC SERIES / PARALLEL - PROBLEM # 4A

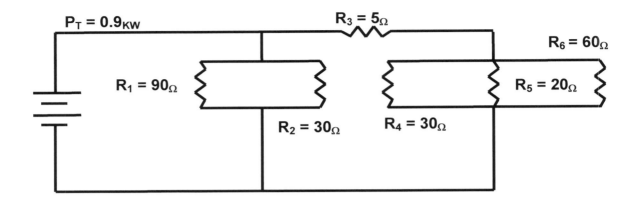

#	R	I	E	P
R_1	90$_\Omega$			
R_2	30$_\Omega$			
R_3	5$_\Omega$			
R_4	30$_\Omega$			
R_5	20$_\Omega$			
R_6	60$_\Omega$			
R_T				900$_W$

DC SERIES / PARALLEL - PROBLEM # 4B

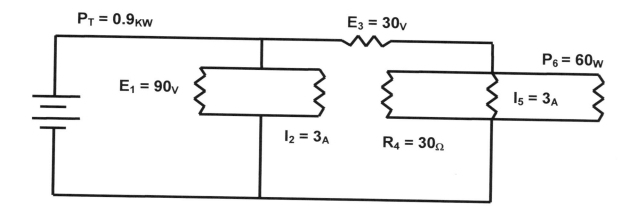

$P_T = 0.9_{KW}$

$E_3 = 30_V$

$P_6 = 60_W$

$E_1 = 90_V$

$I_5 = 3_A$

$I_2 = 3_A$

$R_4 = 30_\Omega$

#	R	I	E	P
R_1			90_V	
R_2		3_A		
R_3			30_V	
R_4	30_Ω			
R_5		3_A		
R_6				60_W
R_T				900_W

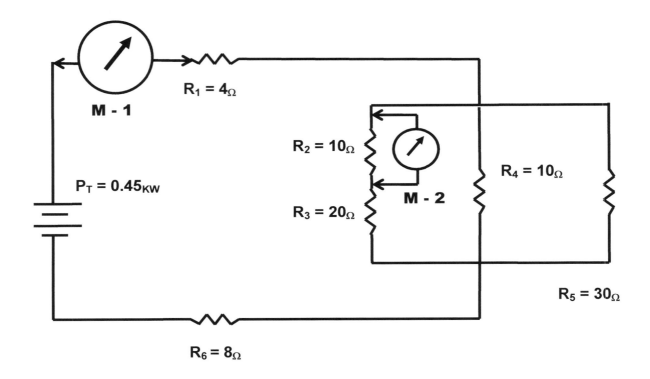

WHAT DOES METER 1 READ AND WHAT TYPE OF METER IS IT?
WHAT DOES METER 2 READ AND WHAT TYPE OF METER IS IT?

#	R	I	E	P
R_1	4_Ω			
R_2	10_Ω			
R_3	20_Ω			
R_4	10_Ω			
R_5	30_Ω			
R_6	8_Ω			
R_T				450_W

DC SERIES / PARALLEL - PROBLEM # 5B

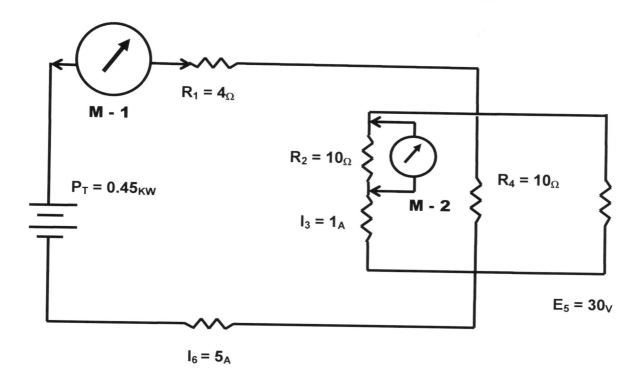

M - 1

$R_1 = 4_\Omega$

$P_T = 0.45_{KW}$

$R_2 = 10_\Omega$

$I_3 = 1_A$

M - 2

$R_4 = 10_\Omega$

$E_5 = 30_V$

$I_6 = 5_A$

WHAT DOES METER 1 READ AND WHAT TYPE OF METER IS IT?
WHAT DOES METER 2 READ AND WHAT TYPE OF METER IS IT?

#	R	I	E	P
R_1	4_Ω			
R_2	10_Ω			
R_3		1_A		
R_4	10_Ω			
R_5			30_V	
R_6		5_A		
R_T				450_W

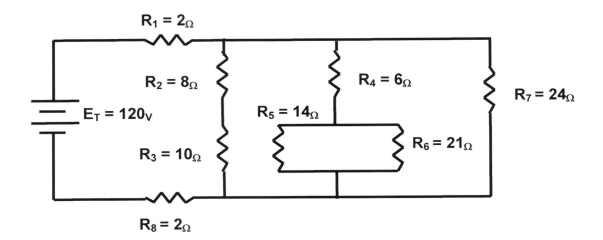

#	R	I	E	P
R_1	2_Ω			
R_2	8_Ω			
R_3	10_Ω			
R_4	6_Ω			
R_5	14_Ω			
R_6	21_Ω			
R_7	24_Ω			
R_8	2_Ω			
R_T			120_V	

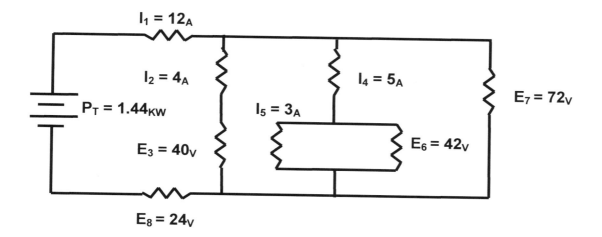

$I_1 = 12_A$

$I_2 = 4_A$

$P_T = 1.44_{KW}$

$I_5 = 3_A$

$I_4 = 5_A$

$E_7 = 72_V$

$E_3 = 40_V$

$E_6 = 42_V$

$E_8 = 24_V$

#	R	I	E	P
R_1		12_A		
R_2		4_A		
R_3			40_V	
R_4		5_A		
R_5		3_A		
R_6			42_V	
R_7			72_V	
R_8			24_V	
R_T				1440_W

DC SERIES / PARALLEL - PROBLEM # 7A

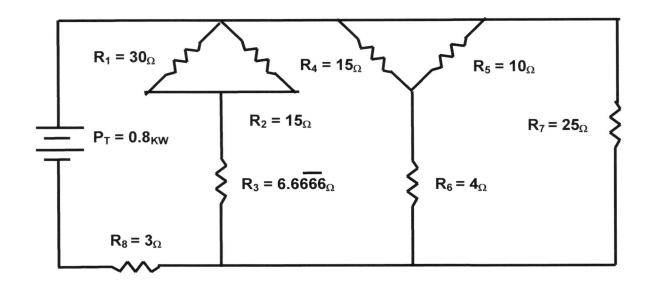

#	R	I	E	P
R_1	30_Ω			
R_2	15_Ω			
R_3	$6.66\overline{66}_\Omega$			
R_4	15_Ω			
R_5	10_Ω			
R_6	4_Ω			
R_7	25_Ω			
R_8	3_Ω			
R_T				800_W

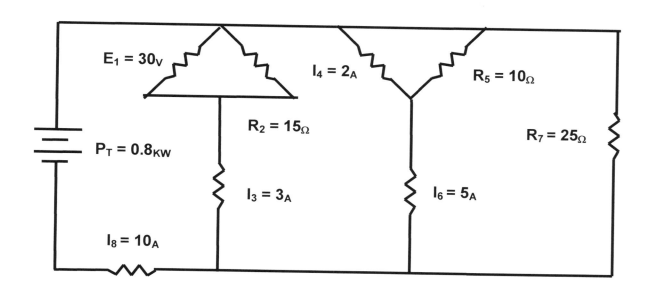

$E_1 = 30_V$

$P_T = 0.8_{KW}$

$I_8 = 10_A$

$R_2 = 15_\Omega$

$I_3 = 3_A$

$I_4 = 2_A$

$I_6 = 5_A$

$R_5 = 10_\Omega$

$R_7 = 25_\Omega$

#	R	I	E	P
R_1			30_V	
R_2	15_Ω			
R_3		3_A		
R_4		2_A		
R_5	10_Ω			
R_6		5_A		
R_7	25_Ω			
R_8		10_A		
R_T				800_W

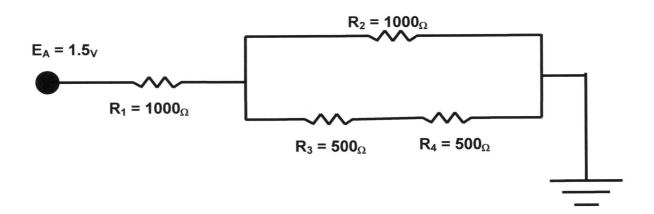

#	R	I	E	P
R_1	1000_Ω			
R_2	1000_Ω			
R_3	500_Ω			
R_4	500_Ω			
R_T			1.5_V	

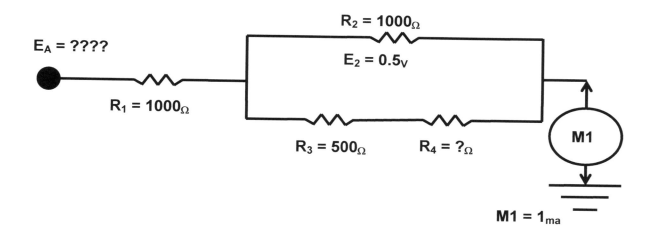

E_A = ????

$R_2 = 1000_\Omega$

$E_2 = 0.5_V$

$R_1 = 1000_\Omega$

$R_3 = 500_\Omega$

$R_4 = ?_\Omega$

M1

M1 = 1_{ma}

WHAT DOES METER 1 READ AND WHAT TYPE OF METER IS IT?

#	R	I	E	P
R_1	1000_Ω			
R_2	1000_Ω		0.5_V	
R_3	500_Ω			
R_4				
R_T		0.001_A		

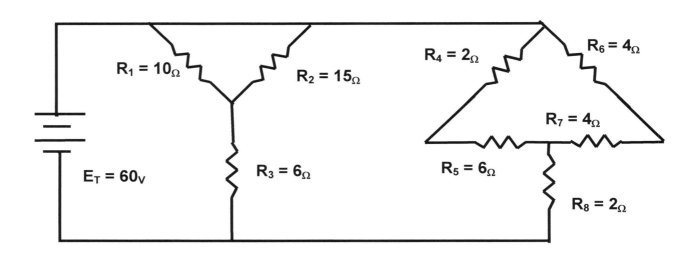

#	R	I	E	P
R_1	10_Ω			
R_2	15_Ω			
R_3	6_Ω			
R_4	2_Ω			
R_5	6_Ω			
R_6	4_Ω			
R_7	4_Ω			
R_8	2_Ω			
R_T			60_V	

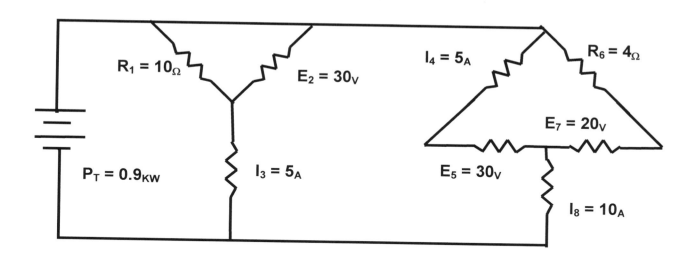

#	R	I	E	P
R_1	10_Ω			
R_2			30_V	
R_3		5_A		
R_4		5_A		
R_5			30_V	
R_6	4_Ω			
R_7			20_V	
R_8		10_A		
R_T				900_W

DC SERIES / PARALLEL - PROBLEM # 10A

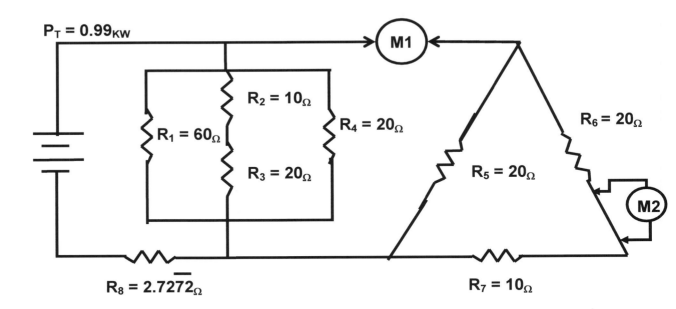

$P_T = 0.99_{KW}$

$R_1 = 60_\Omega$

$R_2 = 10_\Omega$

$R_4 = 20_\Omega$

$R_3 = 20_\Omega$

$R_6 = 20_\Omega$

$R_5 = 20_\Omega$

$R_8 = 2.7\overline{27}2_\Omega$

$R_7 = 10_\Omega$

1) What does meter __ONE__ read and what type of meter is it?

2) What does meter __TWO__ read and what type of meter is it?

#	R	I	E	P
R_1	60_Ω			
R_2	10_Ω			
R_3	20_Ω			
R_4	20_Ω			
R_5	20_Ω			
R_6	20_Ω			
R_7	10_Ω			
R_8	$2.7\overline{27}2_\Omega$			
R_T				990_W

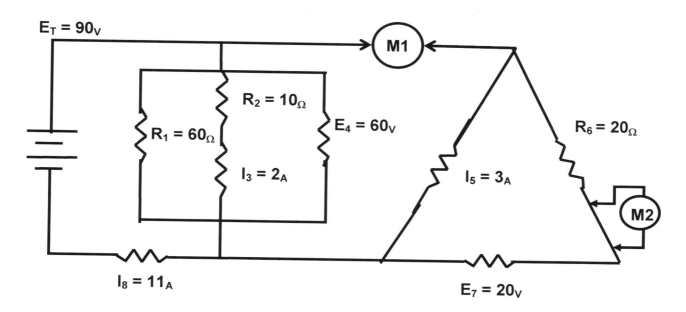

1) What does meter **ONE** read and what type of meter is it?

2) What does meter **TWO** read and what type of meter is it?

#	R	I	E	P
R_1	60_Ω			
R_2	10_Ω			
R_3		2_A		
R_4			60_V	
R_5		3_A		
R_6	20_Ω			
R_7			20_V	
R_8		11_A		
R_T			90_V	

DC SERIES / PARALLEL - PROBLEM # 11A

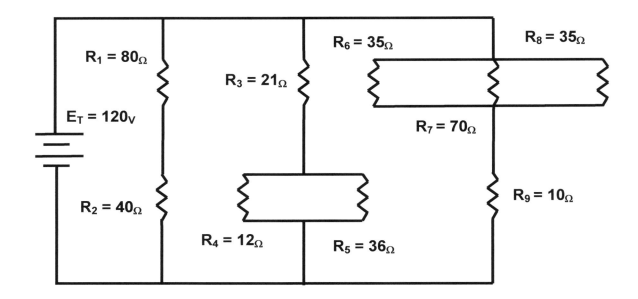

#	R	I	E	P
R_1	80_Ω			
R_2	40_Ω			
R_3	21_Ω			
R_4	12_Ω			
R_5	36_Ω			
R_6	35_Ω			
R_7	70_Ω			
R_8	35_Ω			
R_9	10_Ω			
R_T			120_V	

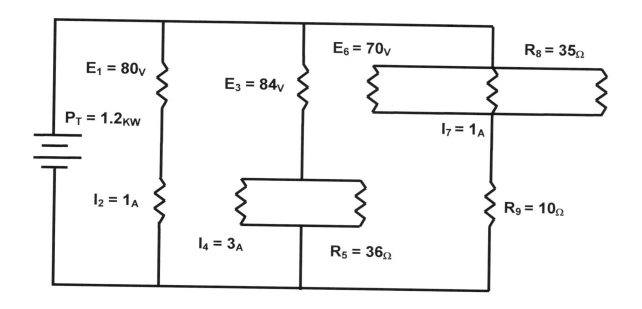

#	R	I	E	P
R_1			80_V	
R_2		1_A		
R_3			84_V	
R_4		3_A		
R_5	36_Ω			
R_6			70_V	
R_7		1_A		
R_8	35_Ω			
R_9	10_Ω			
R_T				1200_W

DC SERIES / PARALLEL - LAB WORKSHEET

INSTRUCTIONS:

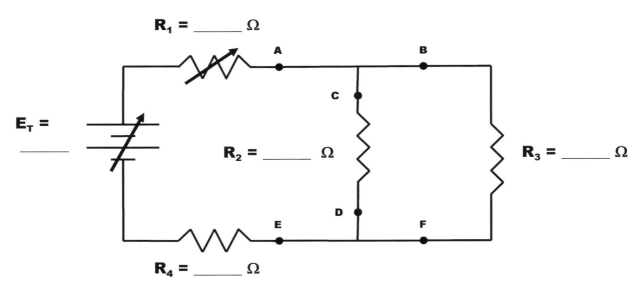

$R_1 =$ _____ Ω

$E_T =$ _____

$R_2 =$ _____ Ω

$R_3 =$ _____ Ω

$R_4 =$ _____ Ω

Figure # 1

1.   With R_1 adjusted to _____ calculate the R_T in the circuit shown in figure # 1.

   $R_T =$ _____

2.   With R_1 adjusted to _____ calculate the R_T in the circuit shown in figure # 1.
   $R_T =$ _____

3. R_T is different between step #1 and step #2. Did R_T go UP / DOWN.
   Explain._____

   _____

4.   With an E_T of _____ calculate the **IR Drop** in step:

   Step #1 _____   Step # 2 _____

5.   Using Kirchhoff's Voltage Law explain the results from step #4.

   _____
   _____
   _____

6.   With the power off, connect the jumpers to construct the circuit in figure # 1.

7.   Call **INSTRUCTOR** prior to proceeding.

8.   Adjust R_1 to the value used in step # 1.

9. Why is there a difference in I_T calculated between step #1 and step #2?

    Explain: _____

    _____

    _____

10. Using a voltmeter, measure the Total **IR Drop** of the circuit in figure # 1: _____

11. Using a voltmeter, measure the **IR Drop** of each resistor in figure # 1.

    E₁ _____   **E₂** _____   **E₃** _____   **E₄** _____

12. Is the value obtained in step # 10, the same as the sum of the voltages in step # 11?
    Explain: _____

    _____

    _____

13. Using the appropriate ammeter function of the multimeter, measure the **I** (current) at points **A** through **F**.

    A. _____   **B.** _____   **C.** _____   **D.** _____   **E.** _____   **F.** _____

14. Adjust **R₁** to the value in step #2, measure the **IR Drop** of each resistor in figure # 1.

    E₁ _____   **E₂** _____   **E₃** _____   **E₄** _____

15. Using the appropriate ammeter function of the multimeter, measure the, **I** (current) at points A through F.

    A. _____   **B.** _____   **C.** _____   **D.** _____   **E.** _____   **F.** _____

16. Define OHM'S LAW. _____

    _____

    _____

17. What is Kirchhoff's Law #1? _____

    _____

    _____

18. What is Kirchhoff's Law #2? _____

    _____

    _____

19. Place ALL equipment in the off position and explain the seven rules to a series-parallel circuit.

    _____

    _____

    _____

    _____

    _____

    _____

$$\alpha + \beta + \gamma = 180^0 \longrightarrow \alpha + \beta + 90^0 = 180^0$$
$$\alpha + \beta = 180^0 - 90^0$$
$$\alpha + \beta = 90^0$$

$$\underline{90^0 - \alpha = \beta}$$

1. $\alpha = 45$ $\beta = $ _____

3. $\alpha = 80$ $\beta = $ _____

5. $\alpha = 75$ $\beta = $ _____

7. $\alpha = 39$ $\beta = $ _____

9. $\alpha = 77$ $\beta = $ _____

11. $\alpha = 45$ $\beta = $ _____

13. $\alpha = 89$ $\beta = $ _____

15. $\alpha = 21$ $\beta = $ _____

17. $\alpha = 53$ $\beta = $ _____

19. $\alpha = 9$ $\beta = $ _____

$$\underline{90^0 - \beta = \alpha}$$

2. $\beta = 45$ $\alpha = $ _____

4. $\beta = 17$ $\alpha = $ _____

6. $\beta = 62$ $\alpha = $ _____

8. $\beta = 31$ $\alpha = $ _____

10. $\beta = 70$ $\alpha = $ _____

12. $\beta = 49$ $\alpha = $ _____

14. $\beta = 9$ $\alpha = $ _____

16. $\beta = 61$ $\alpha = $ _____

18. $\beta = 1$ $\alpha = $ _____

20. $\beta = 69$ $\alpha = $ _____

PYTHAGOREAN THEOREM

$$a^2 + b^2 = c^2$$

$$c = \sqrt{a^2 + b^2} \qquad a = \sqrt{c^2 - b^2} \qquad b = \sqrt{c^2 - a^2}$$

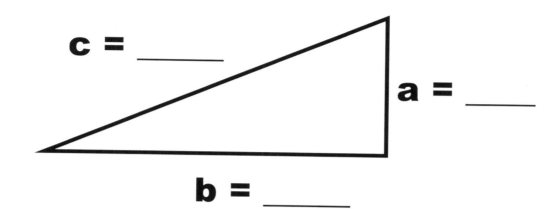

c = _____

a = _____

b = _____

1) a = 3   b = 4   c = _____
2) a = 10   b = 15   c = _____
3) a = 47   b = 35   c = _____
4) a = 20   b = 40   c = _____
5) a = _____   b = 6   c = 10
6) a = _____   b = 85   c = 175
7) a = _____   b = 40   c = 95
8) a = _____   b = 13   c = 19
9) a = 20   b = _____   c = 25
10) a = 30   b = _____   c = 50
11) a = 2200   b = _____   c = 3300
12) a = 300   b = 400   c = _____
13) a = _____   b = 14000   c = 15000
14) a = 500   b = _____   c = 700
15) a = _____   b = _____   c = _____

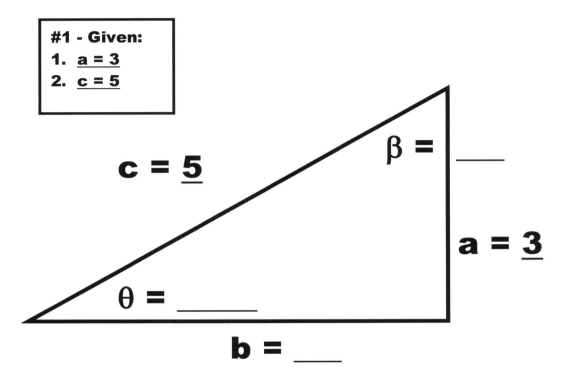

#1 - Given:
1. a = 3
2. c = 5

c = 5

β = _____

a = 3

θ = _____

b = ___

Find:

a = 3

b = _____

c = 5

θ = _____

β = _____

Sin θ (Co-Sin β) = _____

Sin β (Co-Sin θ) = _____

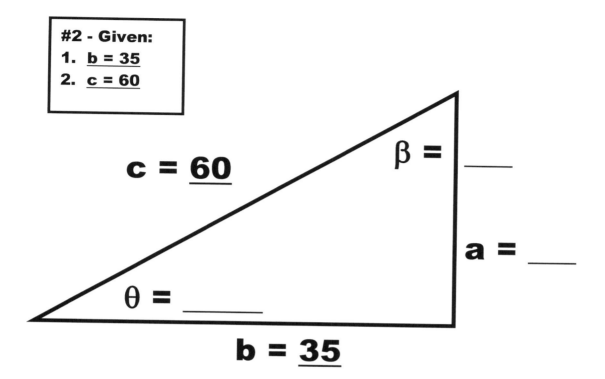

#2 - Given:
1. b = 35
2. c = 60

c = 60

β = ____

a = ___

θ = _____

b = 35

Find:

a = _____
b = 35
c = 60
θ = _____
β = _____
Sin θ (Co-Sin β) = _____
Sin β (Co-Sin θ) = _____

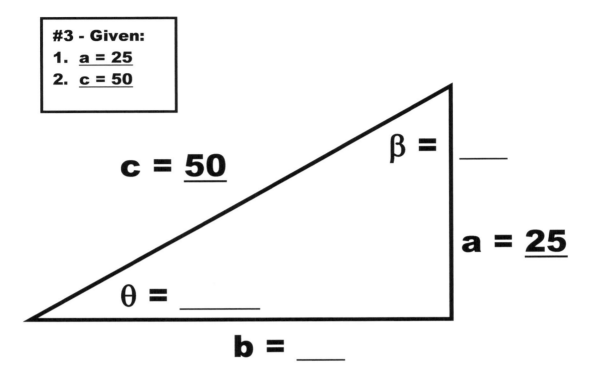

#3 - Given:
1. a = 25
2. c = 50

c = **50**

β = ____

a = **25**

θ = _____

b = ___

<u>Find:</u>

a = **25**

b = _____

c = **50**

θ = _____

β = _____

Sin θ (Co-Sin β) = _____

Sin β (Co-Sin θ) = _____

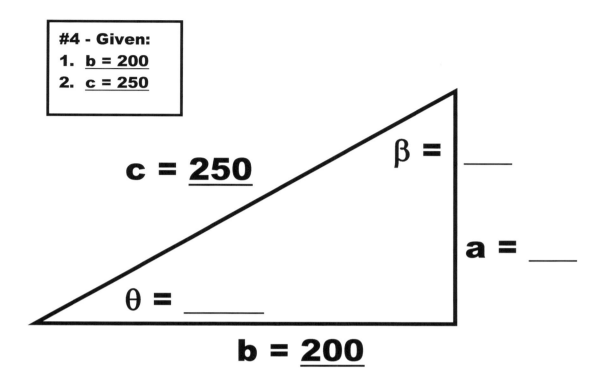

#4 - Given:
1. **b = 200**
2. **c = 250**

c = 250

β = ___

a = ___

θ = ___

b = 200

Find:

a = _____

b = **200**

c = **250**

θ = _____

β = _____

Sin θ (Co-Sin β) = _____

Sin β (Co-Sin θ) = _____

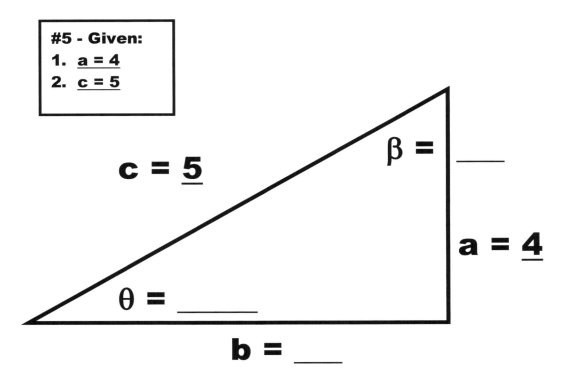

#5 - Given:
1. a = 4
2. c = 5

c = 5

β = ____

a = 4

θ = _____

b = ___

Find:

a = 4

b = _____

c = 5

θ = _____

β = _____

Sin θ (Co-Sin β) = _____

Sin β (Co-Sin θ) = _____

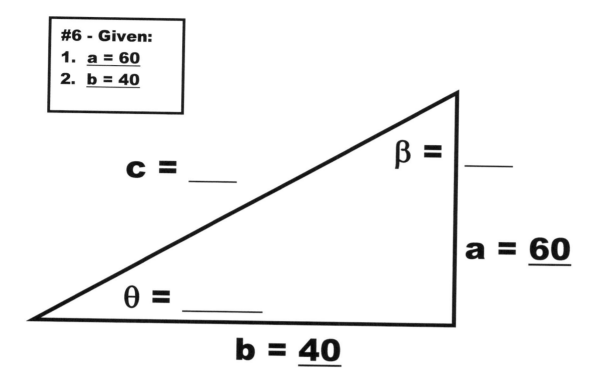

#6 - Given:
1. <u>a = 60</u>
2. <u>b = 40</u>

c = ___

β = ___

a = <u>60</u>

θ = ___

b = <u>40</u>

Find:

a = <u>60</u>

b = <u>40</u>

c = ___

θ = ___

β = ___

Sin θ (Co-Sin β) = ___

Sin β (Co-Sin θ) = ___

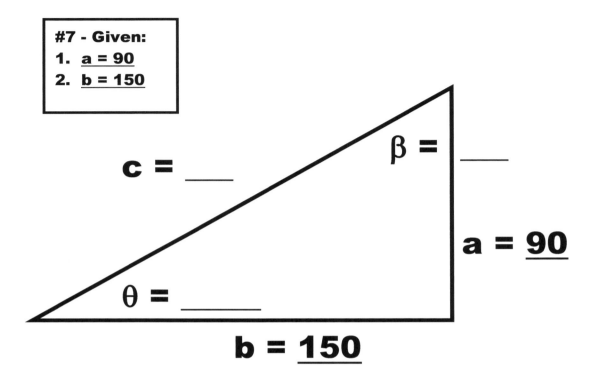

#7 - Given:
1. <u>a = 90</u>
2. <u>b = 150</u>

β = ____

c = ____

a = <u>90</u>

θ = ____

b = <u>150</u>

<u>Find:</u>

a = <u>90</u>

b = <u>150</u>

c = _____

θ = _____

β = _____

Sin θ (Co-Sin β) = _____

Sin β (Co-Sin θ) = _____

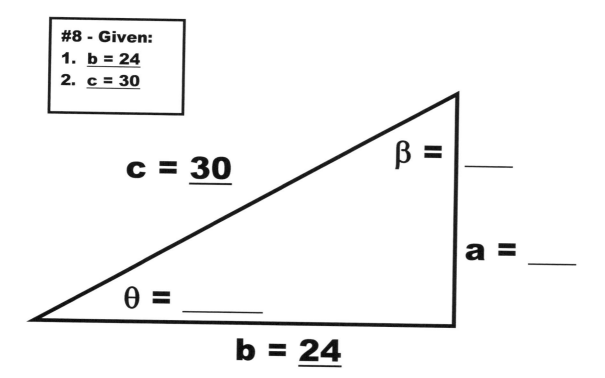

#8 - Given:
1. **b = 24**
2. **c = 30**

c = **30**

β = ____

a = ___

θ = _____

b = **24**

Find:

a = _____

b = **24**

c = **30**

θ = _____

β = _____

Sin θ (Co-Sin β) = _____

Sin β (Co-Sin θ) = _____

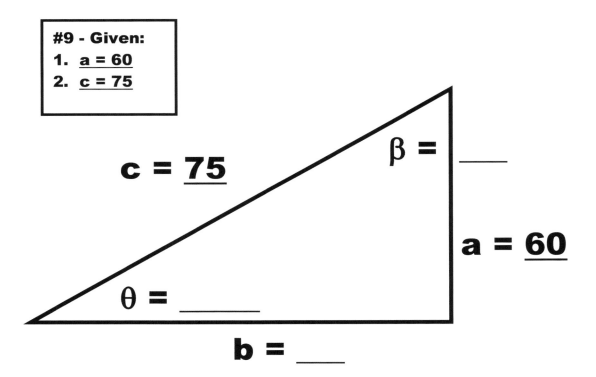

#9 - Given:
1. a = 60
2. c = 75

c = 75

β = ____

a = 60

θ = _____

b = ___

Find:

a =   60

b =   _____

c =   75

θ =   _____

β =   _____

Sin θ (Co-Sin β) = _____

Sin β (Co-Sin θ) = _____

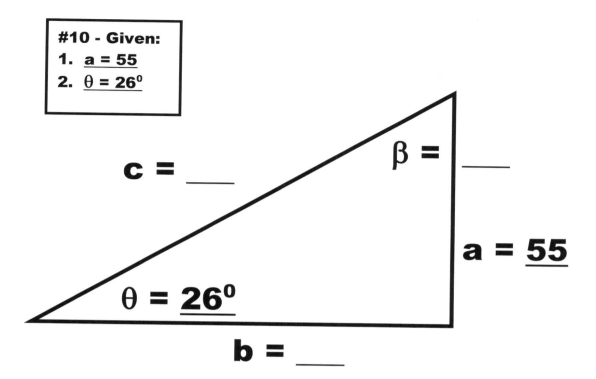

#10 - Given:
1. a = 55
2. $\theta = 26^0$

c = ___

$\beta = $ ___

a = 55

$\theta = 26^0$

b = ___

Find:

a =   55

b =   _____

c =   _____

$\theta = $ 26°

$\beta = $ _____

Sin θ (Co-Sin β) = _____

Sin β (Co-Sin θ) = _____

RC SERIES – PROBLEM # 1

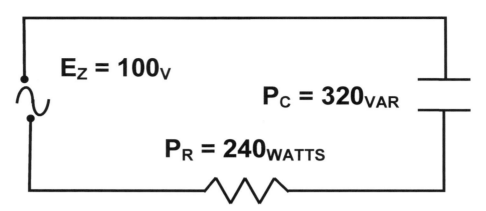

$E_Z = 100_V$

$P_C = 320_{VAR}$

$P_R = 240_{WATTS}$

What is:

p.f. = _____

θ = _____

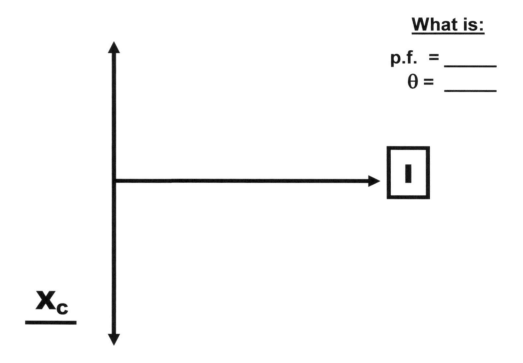

$\underline{X_C}$

#	R	I	E	P
$R_{T \text{ (TOTAL)}}$				240
$X_{C \text{ (TOTAL)}}$				320
$Z_{\text{ (TOTAL)}}$			100	

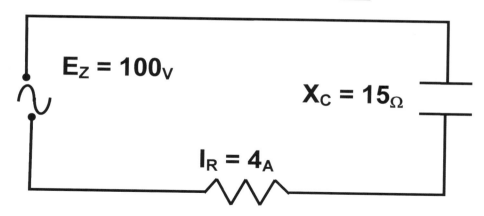

$E_Z = 100_V$

$X_C = 15_\Omega$

$I_R = 4_A$

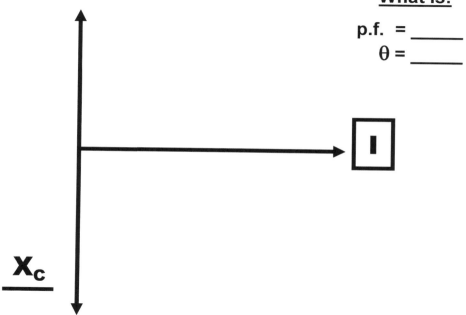

X_c

What is:

p.f. = _____

θ = _____

#	R	I	E	P
R_T		4		
X_c	15			
Z			100	

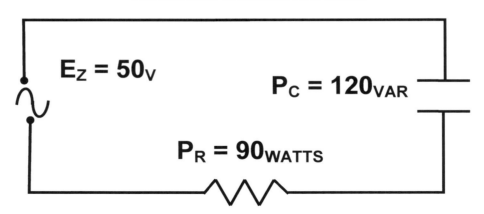

$E_Z = 50_V$

$P_C = 120_{VAR}$

$P_R = 90_{WATTS}$

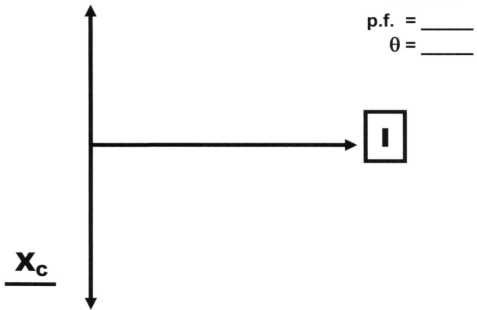

What is:

p.f. = _____

θ = _____

$\underline{\textbf{X}_C}$

#	R	I	E	P
R_T				90
X_C				120
Z			50	

RC PARALLEL – PROBLEM # 1

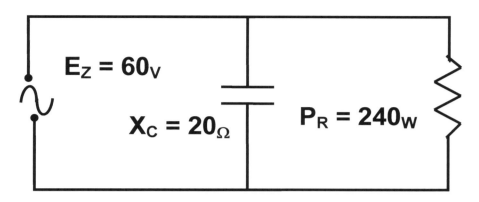

$E_Z = 60_V$

$X_C = 20_\Omega$

$P_R = 240_W$

What is:

p.f. = _____

θ = _____

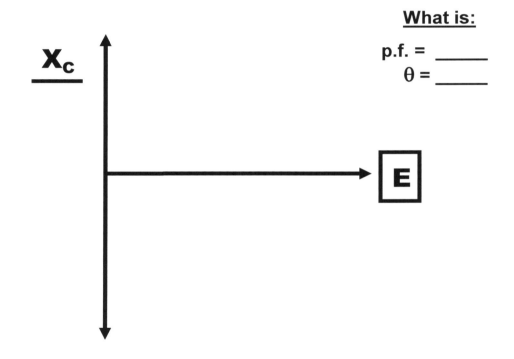

#	R	I	E	P
R_T				240
X_C	20			
Z			60	

RC PARALLEL – PROBLEM # 2

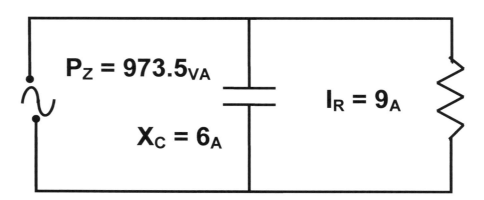

$P_Z = 973.5_{VA}$

$X_C = 6_A$

$I_R = 9_A$

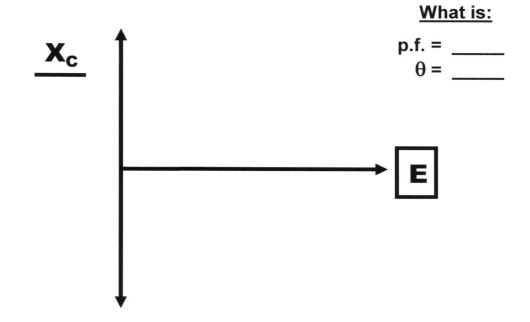

X_c

What is:

p.f. = _____

θ = _____

#	R	I	E	P
R_T		9		
X_c		6		
Z				973.5

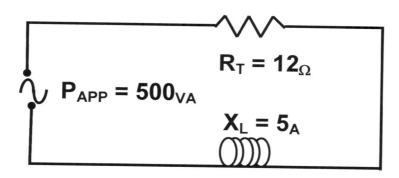

$R_T = 12_\Omega$

$P_{APP} = 500_{VA}$

$X_L = 5_A$

What is:

p.f. = _____

θ = _____

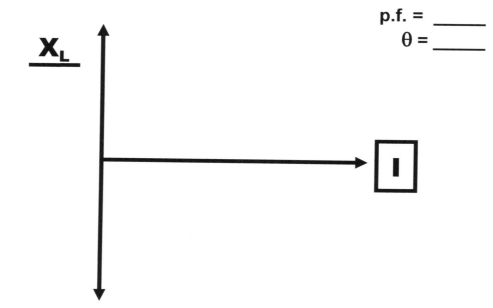

X_L

I

#	R	I	E	P
R_T	12			
X_L		5		
Z				500

RL SERIES – PROBLEM # 2

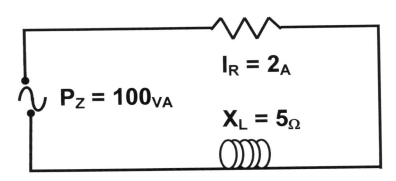

$P_Z = 100_{VA}$

$I_R = 2_A$

$X_L = 5_\Omega$

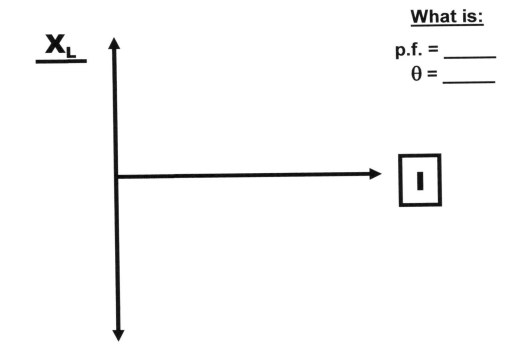

What is:

p.f. = _____

θ = _____

#	R	I	E	P
R_T				
X_L				
Z				

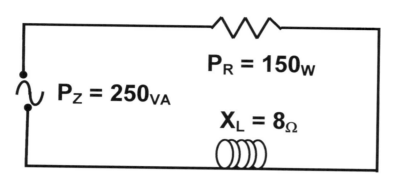

$$P_R = 150_W$$

$$P_Z = 250_{VA}$$

$$X_L = 8_\Omega$$

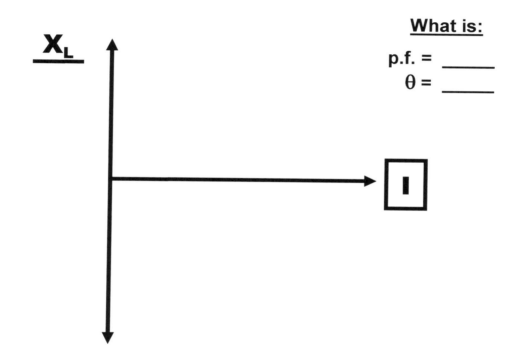

$$\underline{X_L}$$

I

What is:

p.f. = _____

θ = _____

#	R	I	E	P
R_T				150
X_L	8			
Z				250

RL PARALLEL – PROBLEM # 1

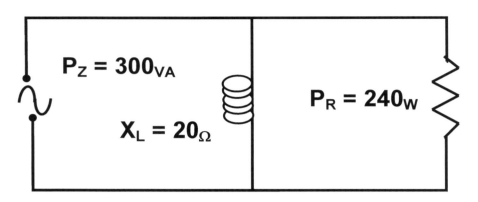

$P_Z = 300_{VA}$

$X_L = 20_\Omega$

$P_R = 240_W$

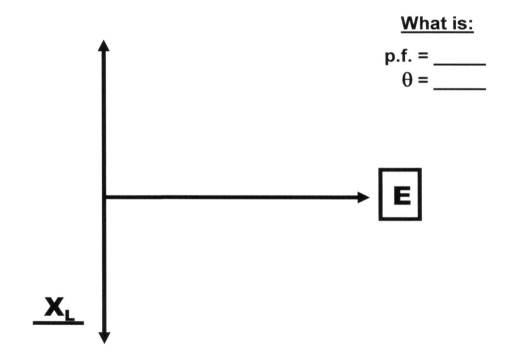

What is:

p.f. = _____

θ = _____

X_L

#	R	I	E	P
R_T				240
X_L	20			
Z				300

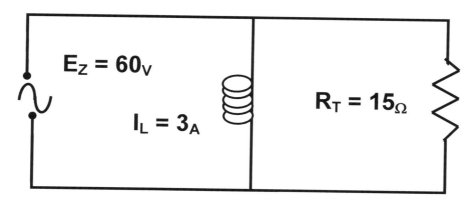

$E_Z = 60_V$

$I_L = 3_A$

$R_T = 15_\Omega$

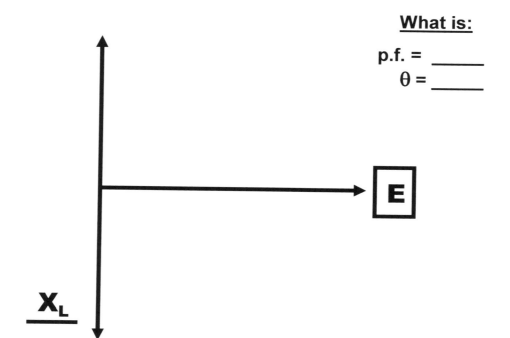

X_L

What is:

p.f. = _____

θ = _____

#	R	I	E	P
R_T	15			
X_L		3		
Z			60	

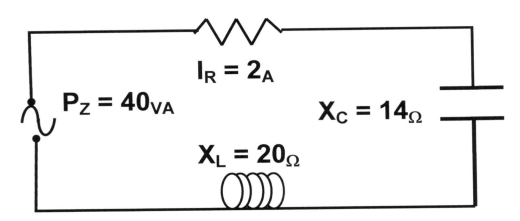

$$I_R = 2_A$$

$$P_Z = 40_{VA}$$

$$X_C = 14_\Omega$$

$$X_L = 20_\Omega$$

What is:

p.f. = _____

θ = _____

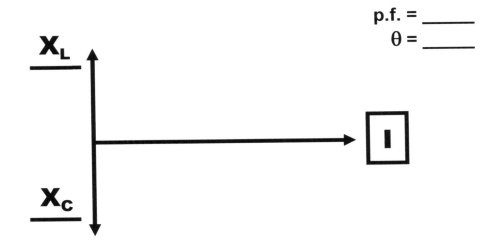

What type of circuit?   A. Resistive   B. Capacitive   C. Inductive

#	R	I	E	P
R_T		2		
X_R				
Z				40
X_C	14			
X_L	20			

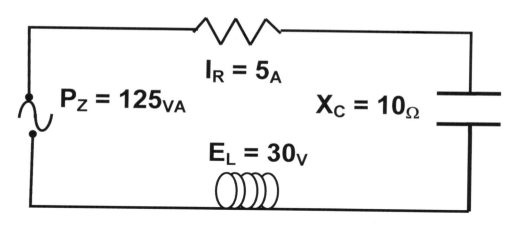

$$I_R = 5_A$$

$$P_Z = 125_{VA}$$

$$X_C = 10_\Omega$$

$$E_L = 30_V$$

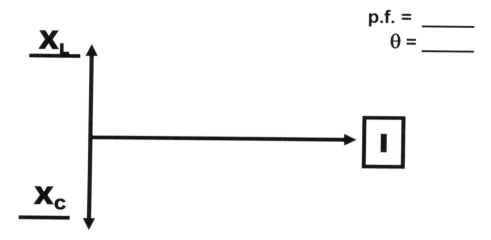

What is:

p.f. = _____

θ = _____

X_L

X_C

I

What type of circuit?   A. Resistive   B. Capacitive   C. Inductive

#	R	I	E	P
R_T		5		
X_R				
Z				125
╳	╳	╳	╳	╳
X_C	10			
X_L			30	

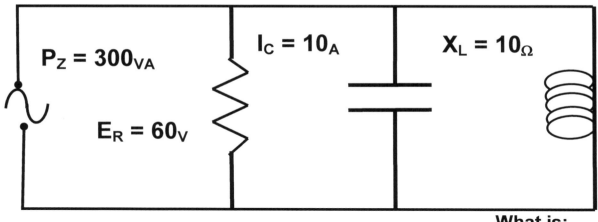

$P_Z = 300_{VA}$

$E_R = 60_V$

$I_C = 10_A$

$X_L = 10_\Omega$

<u>What is:</u>

p.f. = _____

θ = _____

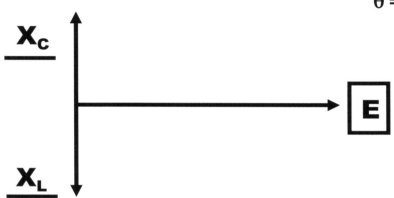

X_C

X_L

E

What type of circuit?   A. Resistive   B. Capacitive   C. Inductive

#	R	I	E	P
R_T			60	
X_R				
Z				300
X_C		10		
X_L	10			

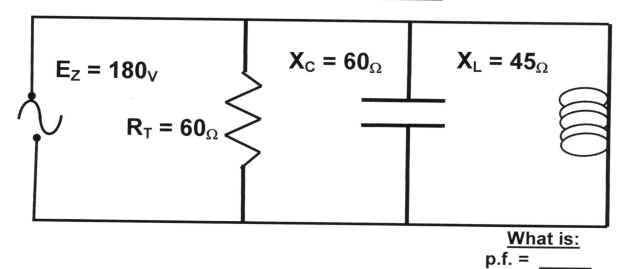

$E_Z = 180_V$

$R_T = 60_\Omega$

$X_C = 60_\Omega$

$X_L = 45_\Omega$

<u>What is:</u>
p.f. = _____
θ = _____

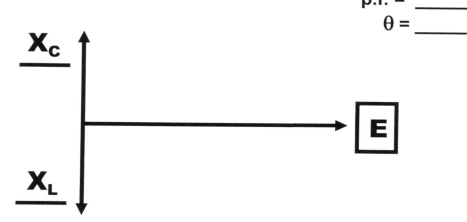

What type of circuit?   A. Resistive   B. Capacitive   C. Inductive

#	R	I	E	P
R_T	60			
X_R				
Z			180	
X_C	60			
X_L	45			

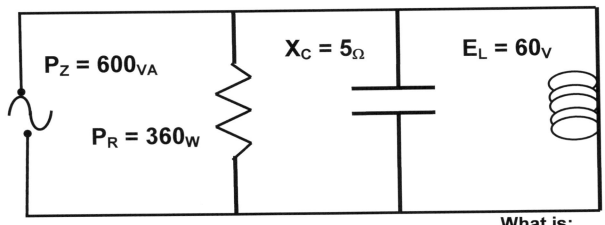

$P_Z = 600_{VA}$

$P_R = 360_W$

$X_C = 5_\Omega$

$E_L = 60_V$

<u>What is:</u>

p.f. = _____

θ = _____

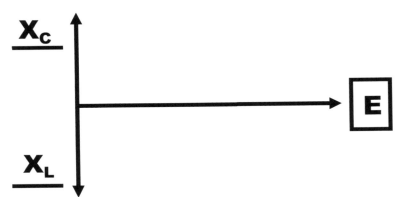

KNOWN: <u>Capacitive</u>

#	R	I	E	P
R_T				360
X_R				
Z				600
X_C	5			
X_L			60	

ALTERNATING CURRENT - LAB WORKSHEET

1. In an AC (alternating current) circuit, the increase in voltage from zero to maximum in one direction, followed by a decrease to zero is referred to as one _____.

2. The effective value of alternating current voltage in a circuit that has a peak voltage of 169.7 volts is _____.

3. A capacitor stores what kind of charge? _____.

4. A capacitor rated at 0.1 uf. and 200 WVDC when fully charged, will have a charge of _____ coulombs. **Q = CE**

5. If a coil is held in the left hand so that the fingers point in the direction of current flow, the thumb will point toward the coil's _____ pole.

6. In a capacitive circuit, the voltage _____ the current by _____ Degrees.   Note: Remember: **ELI** the **ICE** man.

7. Total resistance in an alternating current circuit is referred to as _____Symbol _____.

8. The strength of an electromagnet depends on three variables. They are _____ , _____ , and _____ .

9. The impedance in a series circuit having a resistance of 5 ohms, an inductance of 2.5 mh, a capacitance of 0.5 uf, and a frequency of 400Hz is _____.

10. The frequency at which the circuit described in question 9 will become resonant is _____.

11. In an inductive circuit, when the frequency increases, the inductive reactive _____.

12. In a capacitive circuit, when the capacitance increases, the capacitive reactance _____.

13. A circuit having a 5 uf capacitor in parallel with a 5 mh coil will have an impedance of _____ at 400 Hz.

14. The primary use of a Diode is to: _____
    _____
    _____

15. If a capacitor of 2 uf is connected with a 4 uf in series, across a 120 volt power source, what will the voltage across the capacitors be?

    2 uf = _____ NOTE: $E_C = \dfrac{E_A \times C_T}{C}$

    4 uf = _____

16. A step-up transformer that has a turn's ratio between the primary and secondary windings of 1 : 10 will produce a secondary voltage of _____ when the primary is connected to a 24 VAC source.

17. The effective value of alternating current is the value that will produce the same amount of heat as an equal value of _____ .

18. Sketch, a full-wave rectifier, using four (4) solid state diodes.

19. A 5 Henry coil is in parallel with a 4.5 Henry coil, this will produce a circuit having a total inductance of _____ .

20. 0.4 uf capacitor in series with a 0.5 uf capacitor will produce a circuit that has a X_C of _____ ohms at a frequency of 400 Hz.

21. A circuit having a 5 ohm resistor in parallel with a 5 ohm coil will have an impedance of _____ ohms. NOTE: **BE CAREFUL; Think Twice – Do Once.**

22. Copper loss, eddy currents, hysteresis and flux leakage are losses of power in a _____ .

23. Assuming no loss, a transformer that steps up the voltage from 24 VAC to 240 VAC will produce a secondary current of _____ .

24. List three tests for a capacitor, with an ohmmeter.

    1. _____   2. _____   3. _____

25. What is the effect of connecting a capacitor across the load in a pulsating DC circuit? Explain: _____

    _____

    _____

Made in the USA
San Bernardino, CA
30 August 2016